Searchlight BOOKS

Exploring Robotics

Search and Rescue Robots

Lisa Idzikowski

Lerner Publications ◆ Minneapolis

For my family

Copyright © 2024 by Lerner Publishing Group, Inc.

All rights reserved. International copyright secured. No part of this book may be reproduced, stored in a retrieval system, or transmitted in any form or by any means—electronic, mechanical, photocopying, recording, or otherwise—without the prior written permission of Lerner Publishing Group, Inc., except for the inclusion of brief quotations in an acknowledged review.

Lerner Publications Company
An imprint of Lerner Publishing Group, Inc.
241 First Avenue North
Minneapolis, MN 55401 USA

For reading levels and more information, look up this title at www.lernerbooks.com.

Main body text set in Adrianna Regular.
Typeface provided by Chank.

Editor: Brianna Kaiser

Library of Congress Cataloging-in-Publication Data

Names: Idzikowski, Lisa, author.
Title: Search and rescue robots / Lisa Idzikowski.
Description: Minneapolis : Lerner Publications, [2024] | Series: Searchlight books. Exploring robotics | Includes bibliographical references and index. | Audience: Ages 8–11 | Audience: Grades 4–6 | Summary: "Search and rescue robots assist in looking for people who need help. These robots can reach small spaces, swim, and even fly. Learn about the past, present, and future of search and rescue robots"— Provided by publisher.
Identifiers: LCCN 2022035752 (print) | LCCN 2022035753 (ebook) | ISBN 9781728476803 (library binding) | ISBN 9798765600191 (ebook)
Subjects: LCSH: Robots in search and rescue operations—Juvenile literature.
Classification: LCC TJ211.46 .I39 2024 (print) | LCC TJ211.46 (ebook) | DDC 629.8/92—dc23/eng/20220927

LC record available at https://lccn.loc.gov/2022035752
LC ebook record available at https://lccn.loc.gov/2022035753

Manufactured in the United States of America
1-52261-50701-11/2/2022

Table of Contents

Chapter 1
ROBOTS SEARCH, FIND, AND RESCUE . . . 4

Chapter 2
RESPONDING TO TRAGEDY . . . 10

Chapter 3
TO THE RESCUE . . . 18

Chapter 4
LOOKING AHEAD . . . 23

Glossary • 30
Learn More • 31
Index • 32

Chapter 1

ROBOTS SEARCH, FIND, AND RESCUE

In June 2017, two hikers and their dog were traveling on Devil's Head Trail in Colorado's Pike National Forest. After walking for hours, they lost the trail. They couldn't find their way out of the forest that covered more than 1,720 square miles (4,454 sq. km).

The hikers called 911. A nearby search and rescue team sent an uncrewed aerial vehicle (UAV). The UAV flew over the area, took pictures, and sent them back to the team. The UAV found the hikers in only two hours.

Every year thousands of people become lost, sick, or injured in the outdoors. Others are in danger because of natural disasters. No matter the reason, search and rescue teams are there to help. Technology helps search and rescue teams make quicker and safer rescues.

A remote-controlled search and rescue drone

Why Use Search and Rescue Robots?

During a disaster, search and rescue teams need to act *fast*. Quick action could mean saving a life. Search and rescue robots can accomplish things that people cannot. Robots crawl through rubble into tiny spaces and can go into areas that are unsafe for first responders or search and rescue dogs.

A search and rescue dog helps first responders find missing people after flash floods in Turkey in August 2021.

In March 2022, a search and rescue team works at the site of a plane crash.

Search and rescue robots are machines. People direct some of these robots by remote control. Other machines have computers inside them that guide their actions. Sensors are an important part of every search and rescue robot. Cameras, microphones, and laser scanners allow these machines to locate people who are in trouble and need help. Many search and rescue robots are powered by batteries. Actuators such as treads, motors, and wheels get them moving.

Safety Inspector

Spot, a bright yellow-and-black robot, is about 2 feet (61 cm) tall and 3 feet (91 cm) long. It runs on batteries and remote control. Spot has an important job at a nuclear power plant in South Carolina. It inspects areas in the plant that are hard for humans to access.

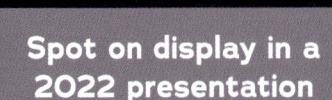

Spot on display in a 2022 presentation

POLICE OFFICERS IN ENGLAND USE A DRONE IN A SEARCH AND RESCUE OPERATION.

▼

Disasters and accidents may happen anytime and anywhere. Roboticists, people who design and build robots, are working to improve search and rescue robots such as Spot to help people in trouble.

Chapter 2

RESPONDING TO TRAGEDY

On September 11, 2001, terrorists hijacked four large planes. The terrorists flew one of the planes into the Pentagon in Arlington, Virginia. The Pentagon is the headquarters of the US Department of Defense. Another plane crashed in a field in Pennsylvania. Terrorists flew two of the planes into the World Trade Center in New York City. Almost three thousand people died in the attacks.

The World Trade Center caught fire and crashed down. Police, firefighters, and other rescue teams rushed to the area. The next morning, Robin Murphy and her team of robotics experts arrived on the scene. They carried special gear—the first ever search and rescue robots.

In 2001, first responders put out fires at the wreckage of the World Trade Center after the attack on September 11.

Building Search and Rescue Robots

In 1995, a bomb exploded in Oklahoma City, Oklahoma. The blast ruined the Alfred P. Murrah Federal Building and nearby buildings and cars. No search and rescue robots existed then. Several roboticists, including Murphy, had an idea. Could robots help search and rescue teams at disaster sites and find buried people? Murphy and her team planned, built, and tested machines.

Investigators search the remains of the Alfred P. Murrah Federal Building.

Six years later, on September 11, Murphy and her team had search and rescue robots ready. A shoebox-sized robot, the micro-VGTV, searched the rubble of the World Trade Center. Treads helped it crawl almost 65 feet (20 m) into the destroyed buildings. The machine inched into places too small for people and dogs to reach. Its cameras sent pictures back to an operator. The remote-controlled robots did not find any survivors, but Murphy and other search and rescue teams learned from the experience.

Workers search the wreckage of the World Trade Center on September 29, 2001.

Key Figure

US computer scientist Robin Murphy is a well-known roboticist. She was one of the first people to create search and rescue robots. Her crew has served at over twenty-five disasters around the world, including the September 11 terrorist attacks and the Fukushima power plant. Murphy has accepted many honors for her work. Her robotics lab is a busy place as new robots are assessed and teams prepare for potential future disasters.

Another Disaster Strikes

On March 11, 2011, a powerful earthquake hit the ocean near Japan, triggering a tsunami with waves up to 33 feet (10 m) high. The natural disasters damaged homes and buildings and caused a crisis at the Fukushima Daiichi Nuclear Power Plant.

Fishing boats release smoke after the March 11, 2011, earthquake and tsunami near Japan.

Workers tried to contain problems at the Fukushima plant, but they failed. The tsunami severely damaged the plant. Surrounding buildings and neighborhoods became unsafe or even deadly because of nuclear radiation released from the power plant. People exposed to unsafe levels of radiation can become sick or die.

▲

AN AERIAL VIEW OF THE FUKUSHIMA PLANT AFTER THE MARCH 11 EARTHQUAKE

The US sent PackBot robots to help. These remote-controlled machines worked inside the ruined plant. They recorded levels of temperature, oxygen, and radiation, and sent pictures back to plant officials.

The Fukushima plant and surrounding areas are still unsafe for people and animals. The air, land, and water contain dangerous radiation. Experts are building and testing new robots to help with cleaning up the nuclear disaster.

The PackBot in 2009

Chapter 3

TO THE RESCUE

Natural disasters are on the rise around the world because of climate change. Roboticists are hurrying to improve robotic technology.

Brothers Mike and Geoff Howe from Maine have built robotic vehicles for years. Their bots are giving firefighters tools to help save lives. The Thermite RS3 is a powerful search and rescue robot. It pumps 2,500 gallons (9,464 L) of water a minute, and one of its cameras senses heat. It also has a plow. Firefighters can use it to push through a wall.

The Thermite RS3 can help firefighters put out fires and push through walls.

By Air

Billions of gallons of burning lava flowed from the active volcano Kīlauea in the Hawaiian Islands in 2018. Search and rescue teams arrived with a pack of UAVs. For the first time, drones worked during a search and rescue mission near a volcano. The DJI 200, 210, Inspire, and Mavic Pro drones made forty-four flights, including night flights. The robots' cameras snapped pictures, and their heat sensors searched for hot spots.

By Land

In 2017, Snakebot moved through the rubble of a collapsed apartment building. It was the robot's first assignment. An earthquake had shaken Mexico City, Mexico, and taken down buildings. The Snakebot's crew wanted to find anyone who was trapped. This 37-inch (94 cm) robot went through tight spaces. Lights and cameras on its head helped users look for people that needed help. The robot didn't find anyone, but it kept search and rescue teams safe.

People remove debris of a building which collapsed after a quake rattled Mexico City on September 19, 2017.

STEM Spotlight

The Defense Advanced Research Projects Agency, or DARPA, works to improve military technology. The DARPA Robotics Challenge, or DRC, tests robots for disaster response. The 2018 DRC targeted robots for use in extreme conditions. Competing robots mapped unknown places and searched for hidden objects inside an underground cave. Two machines controlled by competing teams each found twenty-three of the forty hidden objects. These robots could someday help find people trapped at a disaster site.

By Water

Sometimes, floodwaters wash away buildings and bridges, strand cars on roads, and sweep people into the water. A robotic lifeguard, EMILY (Emergency Integrated Lifesaving Lanyard), has come to the rescue around the world since 2010. This 4-foot (1.2 m) remote-controlled robot swims about 22 miles (35 km) per hour. EMILY moves like a mini jet ski. Up to eight people can grab onto it at a time.

Search and rescue tools, including robots, have improved over the years. But researchers still work to build better technology.

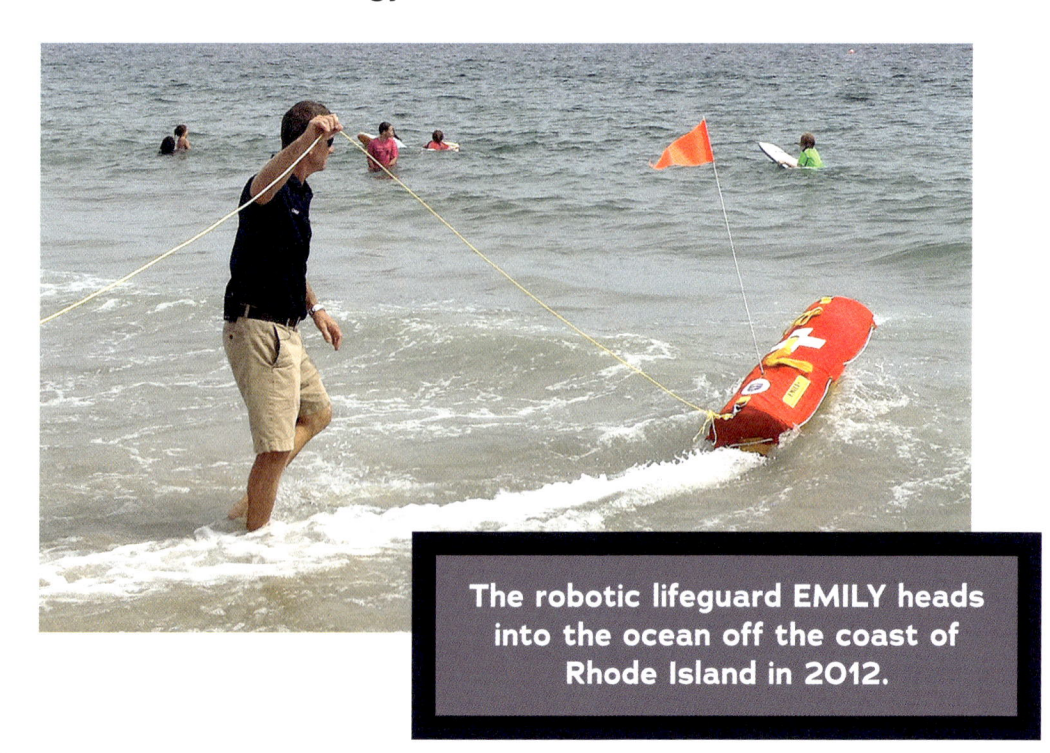

The robotic lifeguard EMILY heads into the ocean off the coast of Rhode Island in 2012.

Chapter 4

LOOKING AHEAD

Roboticists are always busy building and testing new technology. They want to get improved search and rescue robots to teams to save lives.

Roboticists plan robots in two ways. With the top-down method, engineers look at what search and rescue teams want robots to do. Then they build machines to fit those needs. In a bottom-up plan, robots are built and tested first. Then the machines are sent to where they will be useful.

A UNIVERSITY OF CALIFORNIA, BERKELEY, ENGINEERING TEAM'S MODEL OF A ROBOTIC FLY IS SMALLER THAN A US QUARTER.

Robot Designs

RoboFly is slightly larger than a real fly. It can fit into tight spaces that larger drones cannot reach. Laser beams power its wings. During some disasters, gases leak out of damaged buildings. Some of these gases contain chemicals that are unsafe for people to breathe. RoboFly's inventor hopes that these insect-like robots will sniff out gas leaks.

Key Figure

Avye Couloute started writing computer code when she was seven years old. She wanted to get other girls interested in STEM (science, technology, engineering, and mathematics). As a young teen, she started Girls Into Coding. This company gives girls hands-on experience in computing, coding, and robotics skills. In 2022, Couloute became the UK Young Engineer of the Year. She believes that technology can solve global problems.

A girl learns about computer programming in school.

The Heat Assisted Magnetic Recording (HAMR) is a cockroach-like microrobot that jumps, runs, and climbs. For search and rescues, researchers may add heat or chemical sensors, microphones, or cameras to its body. One day, HAMR or similar robots might help at the site of an earthquake, flood, or other disaster.

Search and rescue robots can help teams, such as the one shown here in China in 2022, after an earthquake has hit an area.

SCIENTISTS FIND INSPIRATION FROM ANIMALS SUCH AS FERRETS TO DESIGN ROBOTS.

▼

Scientists are planning to build a ferret-like robot. They want this machine to scoot around like a snake and climb up on rocks. Like the real animal, this robot ferret could stand up on its back legs and look around without tipping over.

THE HUMANOID ROBOT iCUB IN 2020

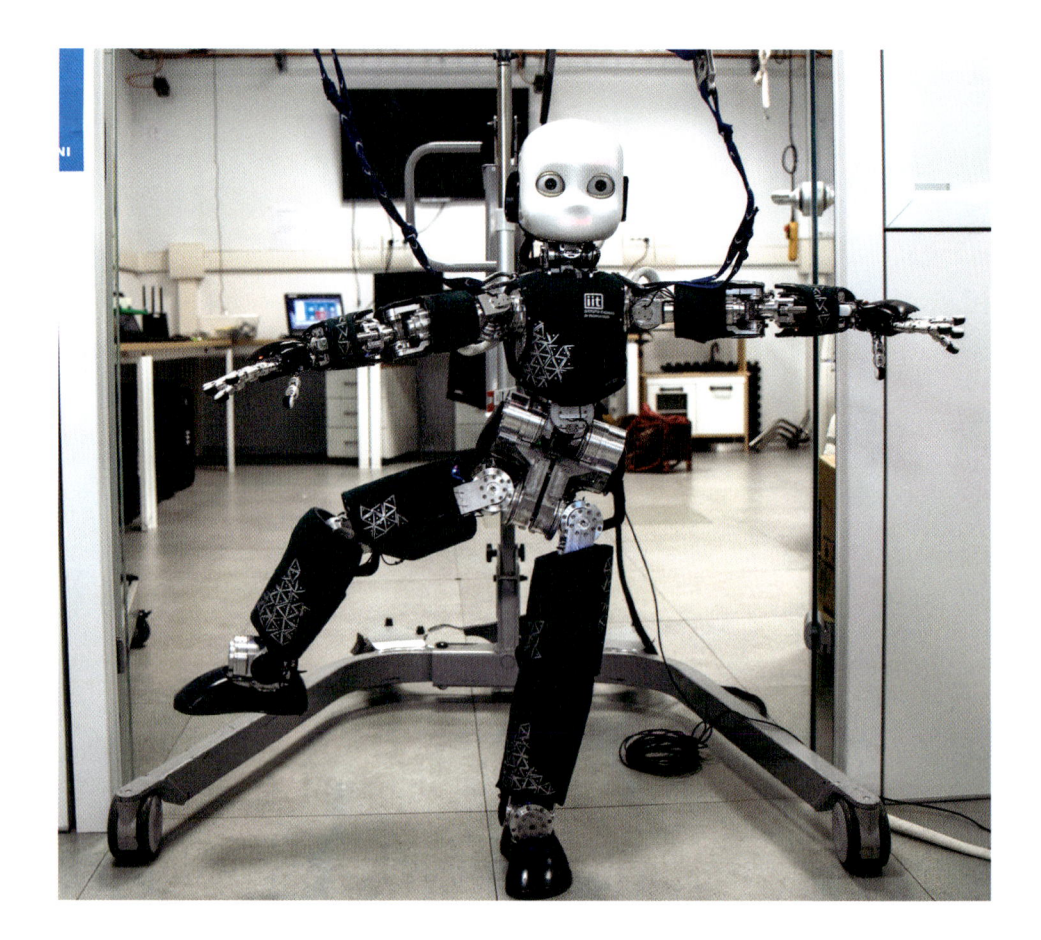

▼

Another group of engineers is creating a flying humanoid search and rescue robot with four jet engines. The team hopes that the iCub and machines like it will someday fly to disaster sites.

Search and rescue teams around the world need support. Improved technology will help them accomplish their goal of supplying aid to people during disasters.

Glossary

actuator: a tool for moving or controlling something

disaster: an event that happens suddenly and causes damage

drone: a flying robot, also known as an uncrewed aerial vehicle (UAV)

first responder: a person who goes to a disaster or accident scene to help

humanoid: a robot with a human form or characteristics

nuclear power: a form of energy

radiation: waves of energy that come from heat, light, or radioactive materials

roboticist: a person who designs and builds robots

sensor: a device that finds and responds to something in an environment

uncrewed aerial vehicle (UAV): an aircraft with no pilot on board